PHYSICAL SCIENCE

FORCES AND CHANGES IN MOTION

By Christina Earley

A Stingray Book

SEAHORSE PUBLISHING

Teaching Tips for Caregivers and Teachers:

This Hi-Lo book features high-interest subject matter that will appeal to all readers in intermediate and middle school grades. It may be enjoyed by students reading at or above grade level as well as by those who are looking for age-appropriate themes matched with a less challenging reading level. Hi-Lo books are ideal for ELL readers, too.

Each book appeals to a striving reader's age and maturity level. Opportunities are provided for students to read words they already know while encountering a limited number of new, high-interest vocabulary words. With these supports in place, students will read more fluently while increasing reading comprehension. Use the following suggestions to help students grow as readers.

- Encourage the student to read independently at home.
- Encourage the student to practice reading aloud.
- Encourage activities that require reading.
- Establish a regular reading time.
- Have the student write questions about what they read.

Teaching Tips for Teachers:

Before Reading

- Ask, "What do I know about this topic?"
- Ask, "What do I want to learn about this topic?"

During Reading

- Ask, "What is the author trying to teach me?"
- Ask, "How is this like something I already know?"

After Reading

- Discuss how the text features (headings, index, etc.) help with understanding the topic.
- Ask, "What interesting or fun fact did you learn?"

TABLE OF CONTENTS

WHAT ARE FORCES?

Forces put things into motion.

A force is a push or a pull.

Some forces are easy to see.

Other forces change how something moves without touching it.

FUN FACTS

The amount of force is measured in units called newtons. They are named after the scientist Sir Isaac Newton.

GRAVITY

Gravity is a force.

Gravity pulls all things with **mass** toward all other things with mass.

It is what holds things to Earth.

Without gravity, everything would float away.

FUN FACTS

Gravity keeps Earth and other planets in our solar system in orbit around the sun.

MAGNETIC FORCE

The force between two **magnets** or a magnet and a metal is magnetic force.

One end of a magnet is the north pole and the other end is the south pole.

Magnetic force pulls the north pole of one magnet to the south pole of another magnet.

FUN FACTS

Some magnets, such as lodestones, are found naturally in the Earth.

ELECTROSTATIC FORCE

Rubbing some materials together can make an electric charge.

This charge moves from one surface to the other.

The charged objects pull or push on other charged objects.

This is electrostatic force.

FORCE OF FRICTION

Friction makes things in **motion** slow down.

If there are no other forces, moving things slow down because of friction.

Smooth surfaces such as ice and glass have less friction.

Rough surfaces such as rock and grass have more friction.

FUN FACTS

We need friction to stand up.

MECHANICAL FORCE

Mechanical forces come from the energy of things in motion.

Cars have mechanical force.

You use this force when your feet push the pedals of a bike.

FUN FACTS

Forces can be balanced or unbalanced. Kicking a soccer ball is an unbalanced force. Floating on top of water is a balanced force.

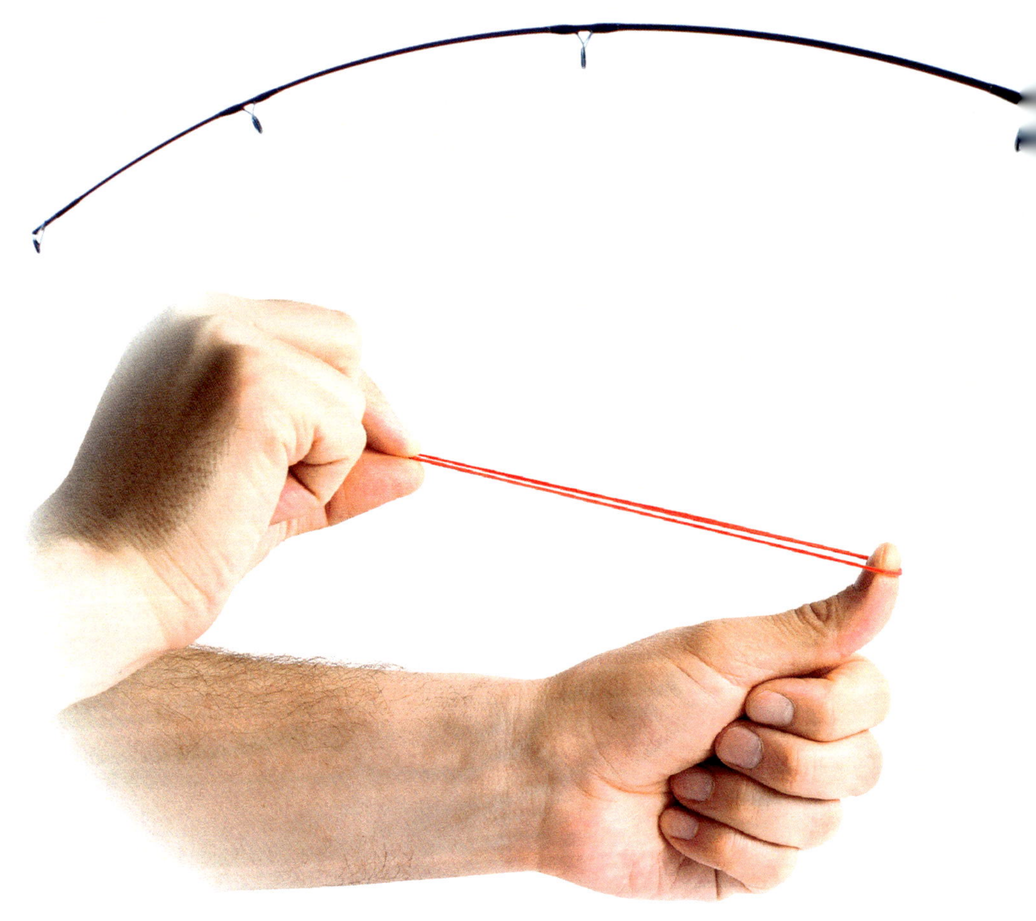

SPRING FORCE

Some objects change their shape when a force is used on them.

When the force stops, the object goes back to the first shape.

Once you stop squeezing a balloon, it returns to its original shape.

Rubber bands, fishing poles, and springs use spring force.

CAREER: TRIBOLOGIST

Tribologists are scientists who study the force of friction.

They work to invent new materials that have less friction and better movement.

These materials can be used for people who need new hip **joints**.

INVESTIGATE: CARS ON ICE

Materials:

- Small toy car (pull-back style or **battery** powered)
- Large baking pan with sides (17 inches by 11 inches)
- Water
- Salt

Procedure:

(1) Fill the baking pan with water and place in freezer overnight.

(2) Launch the car off the ice.

(3) Place the car on the ice. **Observe**.

(4) Sprinkle salt all over the ice.

(5) Launch the car off the ice.

(6) Place the car on the salted ice. Observe.

THE SCIENTIFIC METHOD

- Ask a question.
- Gather information and observe.
- Make a hypothesis or guess the answer.
- Experiment and test your hypothesis, or guess.
- Analyze your test results.
- Modify your hypothesis, if necessary.
- Make a **conclusion**.

SCIENTIST SPOTLIGHT

Wilbur and Orville Wright are known for creating the first flying machine. They wanted to use mechanical force to make a machine lift off the ground. They observed birds in flight. They researched the best places for steady winds. They created a flying machine. After each test flight, they made changes. After three years, Wilbur and Orville Wright succeeded. They created the first flying machine and changed history.

GLOSSARY

battery (BAT·uh·ree): a container consisting of one or more cells in which chemical energy is converted into electricity and used as a source of power

conclusion (kuhn·KLOO·zhuhn): a judgment or decision reached by reasoning

joints (joynts): places where two things or parts, such as bones, are joined

magnets (MAG·nits): pieces of iron that have their component atoms so ordered that the material exhibits properties of magnetism such as attracting other iron pieces

mass (mas): the amount of physical matter that an object contains

motion (MOH·shuhn): movement

observe (uhb·ZURV): to notice something and register it as being significant to use as data

smooth (smooth): flat and even, without indentations

INDEX

AFTER READING QUESTIONS

1. What does gravity do?
2. What are the two poles on a magnet?
3. Which surfaces have a lot of friction?

About the Author

Christina Earley lives in South Florida with her husband, son, and dog. Her favorite subject in school was science. She enjoys learning the science behind the world around her by asking questions such as how do roller coasters work. She loves mint chocolate chip ice cream and mermaids.

Written by: Christina Earley
Design by: Kathy Walsh
Editor: Kim Thompson

Library of Congress PCN Data
Forces and Changes in Motion / Christina Earley
Physical Science
ISBN 978-1-63897-116-0 (hard cover)
ISBN 978-1-63897-202-0 (paperback)
ISBN 978-1-63897-288-4 (EPUB)
ISBN 978-1-63897-374-4 (eBook)
Library of Congress Control Number: 2021945213

Printed in the United States of America.

Photographs/Shutterstock: Cover©oleschwander ©Open Studio, ©Viktoriia Debopre: Cover, Pg 1, 3, 22, 23 ©kentoh: Pg 4 ©Luis Louro: Pg 5 ©Erickson Stock: Pg 6, 7 ©ViktorKozlov: Pg ©Withan Tor: Pg 8 ©WIN12_ET: Pg 10 ©Pau Buera: Pg 10 ©Yulia Nemchenko: Pg 12, 13 ©Lukas Gojda: Pg 14, 15 ©Yaoinlove: Pg 14 ©Red Tiger: Pg 16, 17 ©NATTHAWAT PHROMTHAISONG: Pg 16 ©Sharomka: Pg 18, 19 ©Gorodenkoff: Pg 18 ©ChooChin: Pg 20 ©khuruzero: Pg 20, 21 ©santima studio: Pg 21©Francisco Blanco, ©Pixel-Shot

Seahorse Publishing Company
www.seahorsepub.com

Published in the United States
Seahorse Publishing
PO Box 771325
Coral Springs, FL 33077